Spinning
and
Weaving

Enid Gauldie

N·M·S

NATIONAL MUSEUMS OF SCOTLAND

Front cover: *A detail from* Sunshine, *a painting by Robert Thorburn Ross, 1871.*

Back cover: *Old woman from Skye spinning, with her peat basket on her back and her distaff in her hand, from Kay's* Portraits *about 1812.*

Published by the National Museums of Scotland
Chambers Street, Edinburgh EH1 1JF

ISBN 0948636 68 8

© Enid Gauldie and the Trustees of the National Museums of Scotland 1995

British Library Cataloguing in Publication Data

A catalogue record for this book is available for the British Library

Series editor Iseabail Macleod

Picture research Susan Irvine

Designed and produced by the Publications Office of the National Museums of Scotland

Printed by Ritchie of Edinburgh

Acknowledgements

Front cover: National Galleries of Scotland. 4, 7, 9, 13, 15, 22, 36, 37, ii, iii(right), 42, 45, 46, 50, 54, 56, 58, 61, 65, 67, 68, 70, 72, 73: National Museums of Scotland. 11, 17: by permission of Fife Folk Museum Trust. 14: Mrs Joyce M Sandison (Dr C W Graham's Collection). 20, 78: Ulster Folk and Transport Museum. 25: Reproduced from *Water Power in Scotland, 1550-1870* by kind permission of John Shaw. 29, 53: Collection John R Hume. 30(left), i, iii(left) Scottish National Portrait Gallery. 30(right): New Lanark Conservation Trust. 34: Mary Evans Picture Library. 39: Author. i(top): A detail from a painting by Robert Dodd, 1791, National Maritime Museum London. iv: Anne Campbell, 1 Lickisto, Isle of Harris: Photo Alastair Pout. 48: Crown copyright is reproduced with the permission of the Controller of HMSO. 52: Valentine Collection, St Andrews University Library. 59: Family Archive V W Albrow. 60: Dundee Art Galleries and Museums. 63: From a painting *To Brighton and back for 3s 6d* by Charles Rossiter, 1859, Birmingham City Museums and Art Gallery. 64: by courtesy of the Mitchell Library, Glasgow City Libraries. 68: A Harcus Cutt. 73: J Allan Cash. 74: David Hayes. 75: Norman Burniston at Burniston Studios Ltd.

CONTENTS

INTRODUCTION

Among the flint arrowheads, the pottery shards, the bone pins found on Scottish prehistoric dwelling sites there sometimes appear little round stone discs, about the size of a milk bottle top, with a hole bored through the middle. These are spindle whorls and they are evidence that the people who lived in those houses had the ability to spin yarn and to make cloth, if only by the most primitive and simple methods.

They had little enough to practise on. Scotland was not naturally endowed with the raw materials for making fine cloth. She had no silkworms, no cotton plants, no long-haired goats, no flax. While the tribes at the centre of the earth, the peoples of the Bible who lived around the shores of the Mediterranean Sea, wore soft robes of finely woven linen, when in the Far East the emperors wore beautiful silks and their warriors warm padded cottons, the people of Scotland were still most often clothed in animal skins.

But they began to learn how to shear wool from the fleece of sheep and to wash and comb and spin it into yarn ready for the loom. Successive waves of invaders and adventurers passed through, some of them wearing fine woollen and linen clothes. The indigenous tribes of Scotland began to covet these garments, to recognize their superior comfort. They were ready to acquire knowledge of the techniques which made their manufacture possible.

Wool they already had, even if there was room for vast improvement in the practice of sheep-rearing. Flax had to be introduced. The kind of flax plant suitable for linen production

Spinning with distaff and spindle at Black Corries, Glencoe, 1897. SEA

occurred naturally only in Egypt where it had been extensively and skilfully used for many centuries before its introduction into Europe. But once introduced it proved not too difficult to grow in Scotland even if the quality of the fibre was for a long time poor.

If fur and hide were not immediately relinquished as outer garments, woollen or linen tunics and leggings in imitation of those worn by the incomers were adopted before the period of written history. Fragments of woven woollen and linen cloth have been found along with the other artefacts on archaeological sites. It might be supposed that these had been imported goods were it not for the presence of those little stone discs, the spinning whorls which show that spinning was practised here in prehistoric times.

Through the ages of harsh scarcity and distance from the centres of discovery and ingenuity Scotland took time to emerge into a trading world in which it was possible to exchange the goods and the skills of one country for the products of another. But, in the end, it was the people of this outpost island who introduced to the rest of the world the mechanical inventions of the industrial age.

SPINNING AND WEAVING

1 Prehistoric spinning

Spinning is the process by which fibres are drawn out and twisted together to produce a continuous thread. Very early in human history women discovered that by applying strain to the thread as it spun, a longer, stronger thread could be contrived.

Prehistoric peoples probably first twisted creepers or grasses together to make ropes for tethering animals and tying bundles. From that they gradually learned to produce a finer thread suitable for making clothes. They experimented with different types of fibre, animal hairs combed from domestic beasts, tail hair from hunted animals, rushes from the rivers, long grasses from the plains, even nettles. They used whatever was available to them. They found that by combing and teasing they could produce a soft fibrous mass from which threads could be drawn out and spun.

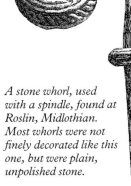

The earliest spinning device was no more than a stick with a flat disc, known as a *whorl*, attached to it for a weight. By drawing out the fibres from the bundle and setting the disc spinning, the thread could be drawn out steadily. The disc was allowed to fall spinning until it reached the ground and the yarn was then wound up on to the stick. This was called *distaff spinning*. Examples of the stone spinning whorl have been found at prehistoric sites in Scotland and can be seen in the Museum. In other places and at other times the whorl might be of clay, metal, bone

A stone whorl, used with a spindle, found at Roslin, Midlothian. Most whorls were not finely decorated like this one, but were plain, unpolished stone.

or wood. Stone was most usual for flax, wood for wool and cotton. The spindles, being of wood, were of course less likely to survive.

Distaff spinning is still used by non-industrialized peoples, such as South American Indians, to this day. It persisted in parts of Scotland into times just within living memory and there are photographs of women spinning on St Kilda and in the Highlands and Islands in the twentieth century.

The distaff was simply a forked stick, about fifteen inches long, onto which the combed fibres were gathered in a tuft, or roughly-formed ball. It was held tucked under the arm or sometimes into a belt worn around the waist. The spinner gradually drew the fibre with her left hand and threaded it through the eye of the spindle which was then allowed to dangle in the air, hanging by the thread from the distaff. It was then given a sharp turn with the right hand until it spun under its own momentum. Gradually, as woodworking skills developed, the rough stick was replaced by a cone-shaped spindle. This gave a more regular spinning motion and produced a smoother thread.

The speed with which yarn could be spun depended on the fineness of the yarn required and the amount of twist needed but, in the hands of a skilled spinner, the quality of the yarn spun by this method was every bit as good as that produced by later inventions. More sophisticated spinning methods might hasten the process but did not necessarily improve the product. The process was easy to learn but hard to perfect. A child could easily be taught to set the whorl spinning and to draw out a thread of sorts, but to do it well, to produce a strong, unknotted thread, took skill and persistence and patience. For the very finest threads it took a spinster working steadily for twelve hours a day to spin only one ounce of yarn. Even for more ordinary yarns it needed the work of several spinners to keep a weaver busy.

All over the world and in every period spinning seems to have been women's work. The Bible describes the virtuous woman 'who seeketh wool and flax and worketh willingly with her hands. She layeth her hands to the spindle and her hands hold the distaff.

All her household are clothed in scarlet.' The task was adopted by women because it could be done while watching a cooking fire or minding babies. The distaff was convenient and light to carry about and could be put down and picked up again without difficulty.

This portability may be one reason why it took so many centuries for this early spinning device to be replaced.

The duty of the daughter living in her father's house was to spin. This is why people writing family history talk about the distaff side for the female line and why unmarried women were, until very recently, always referred to as spinsters. In Scotland, while a well-to-do family might employ servant girls whose chief occupation was spinning, a poor woman and her daughters would labour throughout the hours of daylight to spin enough yarn for the household's needs.

Two hunder year an mair sin syne,
When fashions werna near sae fine,
When common folk had scrimper skill,
And gentles scarce had wealth at will,
When sarks were stark and no that saft,
And lennel worn wi washins aft,
And some had ane, and some had twa,
And mony an ane had nane ava.
When wives wi rocks and spindles span,
And brawest lassies used nae can;
When lassies wi their rocks gaed out
To ane anither nicht about,
A full lang mile o grund an mair,
Sometimes no very free o fear.
On hand reels then they reeled the yarn,
Before the use of wheel or pirn.
But athing has a time atweel,
A time to flourish, time to fail,
So the end of my old tale.

Clothes found on the body of a man in Shetland, about 1700.

This old tale depicts a happy pastoral age when neither rich nor poor were expensively dressed and country girls wandered between each others' houses spinning as they chattered. It has been called the 'kindly society' and while there are aspects that might be envied today, it had its disadvantages. As a wider knowledge spread of how other, perhaps more fortunate people lived there was less likely to be contentment with owning only one shirt and that 'stark and no that saft.' As the desire increased for more varied clothing, for beds and bedding rather than straw and heather, for household textiles of all sorts, the need for spun yarns increased. A faster and more effective method of spinning was needed.

2 Spinning wheels

The first improvement on distaff spinning was the addition of a very simple wheel which drove the spindle round more quickly than could be achieved by the old method. This mechanized the spindle and whorl by mounting the spindle horizontally, where before it had been held upright in the hand, setting it between bearings and grooving the edge of the whorl so that it could act as a pulley. The wheel was revolved by turning with one hand and the other drew out the thread. It was still necessary to interrupt the spin to wind the yarn on to the spindle. Another disadvantage was that, although it was sometimes, in fine weather, lifted outside the cottage door, it was too cumbersome to be carried about by the spinner while she tended her animals. This invention might even be regarded as the first step towards the confinement and restriction of women which only the most recent generations have been spared.

The spindle wheel was unknown in Europe until the thirteenth century and was only gradually adopted thereafter, perhaps not reaching Scotland for another two or three centuries. In the Eastern Highlands no kind of spinning wheel was in general use until the end of the eighteenth century and the distaff was still in use in the West until the second half of the nineteenth.

Simultaneous spinning and winding, without the constant interruptions for winding on, was achieved by the addition of a *flyer*. This kind of spinning wheel begins to appear in illustrations in fifteenth-century Europe. It had on its rotating spindle a bobbin, onto which the thread was wound, and a flyer which revolved at a greater speed than the bobbin and so gave the thread its twist. The fibre, flax or wool, was wrapped loosely round a fixed distaff or *rock* which was set above the spindle. The addition of a treadle allowed the spinner to set the wheel spinning with her foot and keep its action going smoothly while her hands were left free to regulate the flow of thread.

The muckle wheel *in use outside a cottage in Fife, late nineteenth century. Note the yarn winder on the spinner's left.* SEA

This invention was known as the *Saxony wheel* and was quickly taken up throughout Europe although it did not, as we have seen, reach the more isolated parts of Scotland. While the distaff and spindle could be home-produced the new wheel had to be saved for and bought. Resistance to new ideas was reinforced both by poverty and by the need for most women to do more than one thing at one time. The old method remained in use for a very long time. Where this new invention was adopted, however, production of yarn was notably speeded up. Adam Smith, the great Edinburgh economist, suggested that it had doubled productivity in Europe.

This first wheeled device carried only one spindle. In 1764 a wheel carrying two spindles was introduced with which, it was claimed, 'a child may spin twice as much as a grown up person can do with the common wheel'. This was the spinning wheel generally used from then on in Scottish households. It was a handsome piece of furniture, sometimes constructed of fine woods and examples of it are now very much prized as antiques. The frame stood on three feet. On the right-hand side was a spoked wheel, like a bicycle wheel in size but of wood. The treadle turned the wheel by means of a wooden rod connected to the axle of the wheel. On the left hand of the spinner were two spindles, known in Scotland as *pirns*, driven by bands from the wheel. The *distaff* or *rock* with the raw fibre loosely wound round it as before was placed above the pirns and the spinner used both hands to draw out the thread. The spinner moistened the threads continuously with her own saliva to keep the fibre moist and prevent it snapping.

Many spinners suffered from poor health and while we now know that it was caused by poor diet and constant close work in damp cottages the learned people of the day attributed their sickliness to the drying out of their systems by the constant drain of their saliva. 'Spinning, which is the employment of the young women during the winter months, is justly reckoned the cause of consumption among them, by the waste of saliva in that laborious

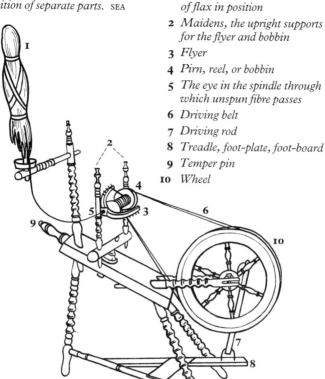

Diagram of flax spinning wheel with definition of separate parts. SEA

1 Distaff shown with a bundle of flax in position
2 Maidens, the upright supports for the flyer and bobbin
3 Flyer
4 Pirn, reel, or bobbin
5 The eye in the spindle through which unspun fibre passes
6 Driving belt
7 Driving rod
8 Treadle, foot-plate, foot-board
9 Temper pin
10 Wheel

exercise,' said one, and another suggested that 'some females are not a little subject to hysterics, a disease attributed partly to the effects of spinning'. By the end of the eighteenth century this two-handed wheel was allowing some women to make eightpence a day, which was twopence more than they could hope to make from field work, the only alternative for most of them. Not surprisingly most families found it necessary to keep the girls at their spinning.

The last improvement on the hand spinning wheel was the introduction of an automatic flyer which saved the time which the spinner had been forced to spend on moving the thread from one wire to another, but this came too late to prevent the inexorable advance of the powered spinning mill and the eventual end of spinning as a domestic occupation.

The quality of the yarn produced was dependent not only on the craftsmanship of the spinner but even more on the preparation of the fibre before it reached her. Spinning, whether by hand or by machine, has to be preceded by effective preparation of raw materials. Careless spinning or poorly prepared fibre resulted in knotted thread which would break easily and give the weavers

Three spinning wheels. There are many variations on the main types of spinning wheel, any of which may appear at different times and in different regions.

Opposite: *Mrs Margaret Taylor of Slamannan, with a variation of the Saxony wheel, 1905.* SEA

Right: *Boudoir wheel, as used by the lady of the* big hoose.

Below: *A smiling Shetland spinner outside her house, 1920s.* SEA

who used it endless trouble and annoyance. This was partly why weavers liked to have their womenfolk providing them with yarn carded and spun at home and under their own eye. Girls were trained by their mothers and even the youngest could be put to winding the yarn ready for the loom.

All the comment of the early eighteenth century reports the poor quality of the flax available to spinners. Cotton has some resilience and was easily adapted first to spinning on the wheel and then to machine spinning. It had been imported from India and had become popular for fashionable ladies' clothes in the seventeenth century but was not at all common in Scotland until after the Union. It began to be used during the second quarter of the eighteenth century but was at first combined with flax to make a *linen union*, cotton not being considered strong enough for the warp. In 1725 the Countess of Leven bought six spindles of cotton yarn from Lady Rossehill and paid four shillings for it. In a private household the Countess, like her friends, could purchase cotton for her own use if she so chose but cotton was at first discouraged in an attempt to protect the linen industry.

Flax preparation gave more trouble. It was known as a greedy crop which starved the land and in the first place farmers had to be persuaded, by example and by the giving of subsidies, to grow more flax so that costs were not kept high by the need to import from Holland, Scotland's chief competitor in the linen markets.

The fibres of the flax plant have to be released from the woody stem of the plant by long processes. First the gathered flax had to be tied in bundles and steeped for weeks in water out of doors to soften it. This is called *retting*. Many of the rectangles of shallow water on farms sometimes mistakenly referred to as duck ponds, were at first flax *stanks*.

After that the bundles had to be taken up and washed and beaten to continue the process of breaking down the fibre. Long before mill spinning, these preparatory processes were mechanized and almost every little river and stream in the Lowlands carried a water-powered mill, variously known as *waulk* mill, *lint* mill, *scutching* mill or *plash* mill, in which flax was prepared for spinning by being beaten and scoured. These little mills were important not only because they improved the quality of flax reaching the spinners but also because those who used them learned to understand and successfully operate water power. They were the forerunners by half a century of industrialization and, in fact, many of their sites, with their dams and lades and water wheels, were later adapted as spinning mills after the introduction of the use of chemicals dramatically speeded-up flax preparation towards the end of the eighteenth century.

After the breaking down of the flax fibre it had still to undergo another process before it reached the spinning wheel. *Heckling* is the means by which the flax mass is combed, straightened,

Harvesting the flax crop in Fife during World War I. Flax had to be pulled by hand, not cut with a bill hook or scythe, because the whole stem was needed.

separated and split into fibres. Beginning as a rough, brown bundle it ends up as a soft, pale, glistening fibre. The hand heckle was a square frame of wood with rows of iron teeth on it which were combed and drawn through and through the flax.

Hecklers were at first independent workers, usually combining together to work in a heckling shed. Before industrialization, when they were relatively well paid, they used to pay one of their number in turns, to read aloud the news sheets or political pamphlets to help pass the time while they worked at what was, unlike weaving, a quiet occupation. This group of men became very articulate and knowledgeable, liable to ask informed and well-prepared questions at public gatherings, which is why *hecklers* at political meetings were given their name.

After mechanization the iron teeth of the *heckle* were attached to cylinders or rollers which were turned by power and the flax was fed into the rollers, often by children. The independent male hecklers, like so many other hand workers in the textile trades, were thrown out of work.

Until the middle of the nineteenth century by far the greatest part of Scotland's population was occupied in either textile manufacture or in agriculture. Before industrialization and town growth the two were often combined, so that part of the day was spent at wheel or loom and part in tilling the land.

So spinning and weaving were, for most people, the two most important tasks outside food production. Clothes, from baby garments to mens' outerwear, were made at home, from flax and wool home grown, home spun and home woven. Everything from infant swaddling to funeral shrouds, literally clothing the family from cradle to grave, was produced either at home or within the immediate community.

Naturally some men and women grew more skilled than others and would exchange the products of their labour for food-stuffs or raw materials. As the more primitive handmade tools gave way, very slowly indeed in some areas, to more sophisticated devices, which were costlier to make, it was more likely that their use would become concentrated in the hands of the best or the most ambitious craftsmen. So while spinning wheels whirred in every cottage and castle a district might contain only one or two weavers. For all of them their tools were important, to be passed on from mother to daughter, from father to son, wheels, heckles and looms a source of income and pride. Any improvements which might be made to them were of interest even if they were not immediately known about or very readily adopted.

3 The Board of Trustees and the spinners

After the Union of the Parliaments in 1707, in an attempt to protect Scotland from the worst effects of competition from its new and more powerful partner, there was set up a remarkable body called the Board of Trustees for Manufactures. Composed largely of lawyers and landowners, this Board was the forerunner of all public bodies established to encourage national economic development. Having funds at its disposal, it set about its task in a very thorough way, first gathering information about the state of industry throughout the country, then making judgements about how it could best be improved and after that offering rewards for improvements achieved.

The Trustees quickly noticed that, if Scottish cloth was going to compete successfully in international markets and especially in the new markets opening up in the Americas, steps had to be taken to improve the quality of the raw materials, which, as we have seen, were often unsatisfactory. They began to import lintseed from the Continent and paid men from the Low Countries to instruct farmers in the best ways of growing flax. One of these men, known as 'Keysar the Flandrican' became a familiar figure riding the clay roads of the Carse of Gowrie.

The Trustees awarded prizes for new tools for heckling and preparing flax for the spinners. The heckles used up to now were very rough, home-made affairs, 'very improper for the purpose, being made with short brass teeth by a sort of Strollers called

Tinkers'. The Trustees sent over to Holland for some samples of the best kind of heckle which they could have imported into Scotland. They paid the salaries of an Englishman and a Dutchman to set up *heckleries* and 'they agreed to breed at the publick expense a certain number of young men in the Business'. This resolution was made in 1734 and was one of their more effective pieces of work because by 1742 they could report that 'The Art of heckling was now much improved and the Trustees began to think of lessening the Expense which they had hitherto bestowed in Breeding people to that Branch of Business.'

Supplied with better flax the Scottish spinners were able to produce a higher quality of yarn. But while spinning for domestic use was very widespread, there were still not nearly enough spinners producing yarn for sale in markets. The Trustees decided to attempt to remedy this by two means. They offered prizes for the inventors of improvements to spinning wheels and they set up spinning schools 'for teaching girls under fourteen years of age to spin'. They were particularly anxious to spread the art of spinning to the Highlands where they reckoned there were too many idle people too easily tempted into rebellion. During 1728 they set up seven spinning schools, three in the Lowlands and four in the Highlands, 'for learning young girls to spin the different grists of yarn, fine and coarse, allowing certain small prizes for those who spun the best and greatest quantities'. In that year they spent £150 on the schools, one of which cost £50, two £30 and four only £10. By 1754 they were supporting nineteen spinning schools at a cost of £371.17.6d.

In 1730 the Trustees paid two Frenchwomen, the wives of weavers already settled by them in Edinburgh, to teach the art of spinning fine yarn fit for cambrics, but in spite of such encouragement and in spite of the fact that some very fine linen was later

Flax heckling. The girl is beating the flax, her companion pulling it through the teeth of heckles attached to a beam.

produced in Scotland, the main trade remained in coarse linens. The Frenchwomen complained that neither the yarn supplied to them nor the wheels provided for them were of the standard to which they were accustomed.

As well as teaching spinning by existing methods the Trustees encouraged invention. In 1747 they announced that 'there were in this year sundry improvements made on the different sorts of spinning wheels whereby more dispatch was made in the work than formerly and those improvements were particularly useful for the coarser kinds of yarns.' In 1750 they spent £621.1.11d on distributing new and improved spinning wheels free to spinners in the Highlands. It would be interesting to know what happened to those wheels.

To make sure that these very large amounts of public money were properly spent they set up a system of inspection and quality control and paid the salaries of yarn sorters and yarn inspectors 'for preventing fraud in the reel and tale'. They took trouble to ensure that poor-quality yarn, or reels of yarn containing fewer yards than they claimed, should not reach the markets and particularly should not be allowed to affect the reputation of Scottish linens in the export markets.

By these means they improved the reputation of Scottish spun yarns both abroad and at home and so increased the demand for them. In spite of their continued efforts they had little success in bringing industry to the Highlands, even after attempting 'to render those employed in the manufacture as independent of the Landowners in that Country as possible'. They were forced to admit that 'the plan had utterly failed'.

But in the Lowlands of Scotland their policies were remarkably successful. As demand increased, the price spinners could ask rose and a certain measure of prosperity reached country

A modern spinning-wheel maker, Donald
Sinclair, at Port Sonachan, Loch Awe.
Alasdair Alpin MacGregor, SEA

homes. By 1749 the Trustees considered the price of spinning 'by much too high'. In 1747 the weavers were complaining 'that great quantities of the yarn of this country continued to be carried to England and Ireland, by which means the price was raised upon them but as this contributed to spread the skill of spinning over the country the Trustees did not so much regret it'.

These improvements in the quality of the spinners' work and in the reward to be gained from it were, however, short-lived. Even as the Trustees recorded the improvements in the linen trade, the cotton industry in the north of England was making advances which were shortly to be copied in Scotland and which were to reduce the status of the spinners to abysmal levels of poverty, discomfort and degradation.

4 The putting-out system

Until the eighteenth century the biggest part of the yarn spun at home was needed for domestic use. After the loom had been supplied, the family clothed and the home provided for, any surplus might go to market, but families found a use for most of their own product.

During the eighteenth century, market demand and speeded-up processes, plus a slightly easier flow of capital, stimulated the development of the *'putting out' system*. Spinners still worked at home, at their own wheels and in their own time, but they no longer owned the materials on which they worked. The habit of growing a row or two of flax among the food crops died out. Now dealers, travelling about the country, bought flax on a larger scale from farmers, farmers who were paid premiums for growing flax by the Board of Trustees and were glad to have it profitably sold.

The dealers then arranged for the treating and preparing of the field crop, sometimes at their own lint mills, sometimes by arrangement with bleachfields, before distributing bundles of flax to the domestic spinners. These were men who could afford to take the risk of lying out of capital between the buying of the raw flax and the selling of spun yarn or finished web and who had, besides, money to run fleets of carts and horses. Spinners were paid for their work which was, when spun and reeled, collected in the dealers' carts for distribution to weavers and bleachers. They lost the status of the self-employed, the independent worker, and became tied to the dealers' goodwill.

Prices were not standard. A good spinner, capable of producing an even, fine yarn, was paid a higher wage than the less efficient. But because the dealer was in a position to set the prices and could leave unfavoured spinners without yarn to work on, he could withhold payment if he found the slightest fault or variation in quality. The spinners were always poor and often in his debt. Sometimes he would supply them with foodstuffs on credit, taking back the price out of their wages so that they could find themselves working for nothing.

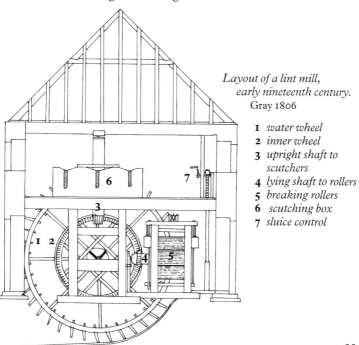

Layout of a lint mill, early nineteenth century.
Gray 1806

1 *water wheel*
2 *inner wheel*
3 *upright shaft to scutchers*
4 *lying shaft to rollers*
5 *breaking rollers*
6 *scutching box*
7 *sluice control*

He always weighed the flax he put out to the spinner and weighed it again when he collected the spun yarn. There was bound to be some loss in weight during spinning, especially if the flax was of poor quality with many impurities to cause breakages and knotting, but the dealer was always on the watch for spinners who might try to keep back some of his flax for their own use. If he suspected, however unjustly, that they were cheating him, he could hold back wages, or even worse, fail to supply them with more flax to spin.

Some dealers, beginning in a small way and usually buying on credit, became rich and powerful by these means. Because they soon had all branches of the textile trade under their control, they could organize it to their own advantage. The spinners became very dependent on them, both for materials to work upon and for the money wages that now came into the households. It was often from this class of business man that the large-scale entrepreneurs of the industrial revolution arose, men who had acquired both capital and knowledge of the industry through years of travelling and dealing. If such a man set up a mill in his district and ceased to put out flax to the home spinners they were left with very little alternative but to come into the mill to work for wages.

Linen dealers, hard though they perhaps seemed in their dealings with the spinners, were important in preparing Scotland's textile industry for the changes that were to come. Some of them were kindly enough men whose visits were welcome to people who saw very little other commerce with the outside world. They did bring small money wages into country households which had hardly known such a thing and so, in a minor way and for a limited time, brought a measure of purchasing power into rural areas. They did spread what the eighteenth century was so fond of calling 'the Spirit of Enterprise' throughout the country. And, because of their power to withhold wages or to refuse employment for poor quality, they did raise the standard of yarn supplied to the markets so that the custom for Scottish yarn was greatly increased.

These dealers were Lowland men, dealing in flax for the linen industry and, although pushing their trade further into the hill areas than anyone had gone before, they were still involved to a large extent with the hinterland of the Lowland cities. In the woollen trade of the Highland areas until the nineteenth century, the degree of organization depended upon the needs and the ambitions of local weavers, many of whom used only the products of their own flocks and the work of their own womenfolk, although some of them may have bought fleeces and put out spinning to local women.

The cotton trade was different again. In the West of Scotland, dealers looking for spun cotton to supply the weavers of cotton handkerchiefs, muslins and cambrics, found it better to import cotton yarn from England than to develop a system of putting out cotton to local spinners. Because it proved easier to adapt cotton to machine spinning, the cotton industry in Scotland was very quickly industrialized and did not need the long period of country dealing through which the linen industry evolved.

5 Water-powered spinning

In 1733 in Lancashire the invention of John Kay's flying shuttle allowed weavers to work much faster, so using much more yarn in a given period. The cry of the weaver at home was for 'mair pirns, mair pirns' and his wife was harassed by the need to supply them. The balance of work between weaver and spinner had been disturbed and this meant that for a time the work of the spinner was in great demand.

Naturally, where the need was so obvious powers of invention were stimulated. But it was some time before a major improvement was made to the spinning wheel which was, in its way, a thing of great perfection. In the 1770s James Hargreaves, another Lancashire man, with nothing more than a penknife to help him, produced his *spinning jenny*. A crude but useful machine, the jenny could turn eight spindles where the old wheel turned two and before long, with only minor adjustments, it could turn 120 spindles. This was still a hand machine, one which could be used at home by women and girls, but it speeded production to the point where the balance between spinner and weaver was once again thrown out of kilter and the spinner's work devalued. Its success attracted great hostility and many of Hargreaves' early machines were smashed.

Also in the north of England, the inventions of Richard Arkwright and Samuel Crompton allowed the application of power to spinning machinery and cleared the way for the change from small-scale domestic occupation to industrialized factory production. It was the scarcity of yarn that directed attention to the need for innovation in the spinning process. Now the market was to be flooded with cheap yarns in such quantities that the weavers could not use them fast enough.

Richard Arkwright was a barber who had built up a successful business dealing in hair for wigs. He saw the need for a machine which could produce a cotton yarn strong enough for warp thread, which up to then had been spun from linen. After many attempts he invented a spinning frame which could be turned by water. This was so ingenious that it was quickly copied and Arkwright had difficulty in protecting his patent.

The next improvement was Samuel Crompton's *spinning mule*, so called because it combined the benefits of Hargreaves' jenny with Arkwright's water frame. These inventions aroused envious imitation from capitalists and great hostility from the working spinners, who feared they would be robbed of their employment. Development of these new machines took place in the north of England because large amounts of capital had already been invested in the cotton trade there. But it was not long before the demand for cotton goods in America and the availability of capital from the tobacco trade in Glasgow led to the setting up of cotton mills in the West of Scotland.

The first water-powered cotton mill in Scotland was operating in Penicuik, Midlothian, by 1778. By the 1780s spinning mills had been set going at Paisley, East Kilbride, Woodside in Glasgow and Stanley in Perthshire. By 1814 there were 120 cotton mills in Scotland, by far the greatest number of them in Lanarkshire and Renfrewshire.

The first mills were small-scale affairs, but when Richard Arkwright visited Glasgow and met David Dale, a rich importer of continental yarns, the stage was set for large-scale expansion. In the New Lanark mills on the Clyde, later to be managed by Dale's son-in-law Robert Owen, Scotland was to see not only the employment of thousands of people in one enterprise but also a social experiment. Owen, seeing himself as a benevolent philanthropist, directed every aspect of the lives of 'the little elves' he employed, some of them brought to Dale's mills from workhouses and orphanages and so without any one else to protect

New Lanark Mills, 1817, showing the scale of Owen's social experiment.

David Dale, founder of New Lanark and father-in-law of Robert Owen.

New Lanark mill-worker Margaret Stewart on the steps of her house in Caithness Row, about 1905.

them. If his paternalism is less highly regarded now than it was in his own time, it was followed by men of much less vision, less benevolence and more greed and the cotton mills in Western Scotland were to produce conditions of cruelty and misery which, when at last they were acknowledged, were to appal the civilized world. The very success of the mills and the rate at which they spun out hitherto unimaginable quantities of yarn kept prices low and wages lower. Adults could earn so little that they were forced to take even the smallest children of the family into the mills with them because the tiny amount those children could earn was necessary to supply the family's needs.

Before the application of power to spinning, although they were accustomed to work long hours, workers had been able to take a break now and then, to stretch weary limbs, step outside to breathe the air, allow themselves a drink of water and particularly to begin and end the working day when it suited them. Even for hard-working mothers it was possible to vary the demands made upon them, a little while in the kitchen, a little while in the kail-yard, a little while at the wheel. And in more prosperous times the whole family would allow itself an occasional day off work.

Water-powered spinning mills did not immediately or in all districts drive the hand spinners out of existence but it was not long before the domestic trade became quite unable to compete and the reluctant spinners were forced to seek employment in the new mills. There they were forced to begin at a set hour, stay at their work all day until released by the overseer and attend every day and all day to one boring and repetitive task. They found it hard. The water-powered mills were, for the most part, country mills, drawing their work force from a rural population. People who had been used to living close to nature found relent-less indoor labour hard to get used to. 'As soon put a deer to the plough as a Highlander to the loom' said one report. In 1833 mill spinners worked from 5.30 in the morning until seven at night with only two half-hour intervals for meals. Children fell asleep at their work and young women fainted with exhaustion.

Water mills were dependent for their power on the supply of water to the wheel. In periods of drought the streams which fed the dams sometimes dried up and there was not sufficient force of water to turn the wheel. Then the hands would be laid off and unpaid. But when the dam filled again masters anxious to fill orders would have their workers roused from their beds, often in the middle of the night, to take advantage of the flow of water and keep the machines turning.

Flax did not lend itself as successfully as cotton to mill spinning and many unsuccessful attempts were made to set up flax-spin-ning mills. John Kendrew and Thomas Porthouse of Darlington invented a dry flax-spinning machine in 1787. In the same year the first flax-spinning mill in Scotland was set up on the banks of the Bervie Water in Kincardineshire. In 1790 another flax-spinning mill, this time for the heavier yarns used in Dundee's coarse linen trade, was started at Kinnettles, near Forfar. From that year onwards the linen mills in the East of Scotland proliferated as the cotton mills had already done in the West. The first flax-spinning mill in Dundee itself was started at the Chapelshade in 1792 with an old sun-and-planet steam engine made by James Watt.

Flax needs to be kept moist if it is not to snap and flax-spinning machinery was not really successful at first. Until the machinery was properly understood there were many breakdowns. The engineers who installed the Chapelshade machinery used to tell how, when the frames were first started, the rattle and strain of the machinery were so great that they used to flee outside and peer through the window to watch the breaking pieces being flung about the mill floor. Damage was often so severe that it took weeks for repairs before spinning could begin again. The men with capital invested in the business were so anxious for success that they attempted to push mechanization along in imitation of the cotton spinners before all the problems had been properly worked through.

The wet-spinning frame solved most of the problems. The flax, instead of being fed straight into the rollers, was passed through a trough filled with water which was kept at a temperature of 150% to 180% by steam pipes passing through the trough. The fibres passed slowly through this hot water, became macerated and soft to a state something like soft indiarubber. From there it passed to the drawing rollers which were made of brass and very finely fluted. But if this machine improved the spinning it did not help the spinners. The hot water and steam spray drenched and sometimes scalded them as they worked. 'Wet spinning is certainly the least healthful branch of the manufactory .... I witnessed a more painful sight again and again in beholding the miserable, unhealthy looking beings in the wet spinning department than in any other parts of the many manufactories I have now visited.'

6 Steam power

The need for water power had placed the early spinning mills along the streams in country areas. The growth of population in towns had left many of them without sufficient water for ordinary domestic purposes and certainly without enough to provide power for the water wheels. So the mills were begun away from the centres of population, where there was no captive labour force.

Very often the need to attract labour to the mills had encouraged landowners or mill masters to build cottages, sometimes whole villages, with churches and schools for their workers. High ideals often characterized the talk of these early entrepreneurs and they prided themselves on giving employment, on regularizing the lives of people upon whose previous lazy undisciplined lives they frowned. In fact reality was often very different from the vision and the lives of the workers were both harsh and dangerous. Damp conditions, long hours and severe overseers affected their health and the unfenced machinery caused many terrible accidents.

But if life in the water-powered mills was hard it was nothing to what was to come when steam arrived. The application of steam power to spinning machinery required the expenditure of such large amounts of money that the mill owners felt pressed to get maximum returns for their outlay. This meant keeping the machinery going ceaselessly. Hours when the machines lay idle seemed to the mill owners like deliberately thrown-away money, and so they were kept working day and night the year round. With no regulation of working hours men, women and children were driven to tend those machines for lengths of time which now seem unimaginable. And the pace set by the steam engine which drove the machines had to be maintained. The workers had for their own safety to pay close attention to their work and could never step aside from it for a moment. Overseers walked the spinning floors with tawse and whistle in hand to keep control of the very young children who worked between the spinners' feet, shifting bobbins and removing gathered fluff under the spinning frames.

The masters actually preferred to employ very small children for this job because they could squeeze under the machines to clean without the need to stop the engines first.

Because a steam engine could easily provide the power to turn thousands of spindles, these new steam-powered mills were built on an enormous scale, employing not hundreds but thousands of workers. This meant that the worker was even further separated from the man who employed him and it was perfectly possible for mill owners and their families, growing wealthy through the labour of the people in their mills, to have no knowledge of the conditions under which those people worked and lived. They could truly believe themselves to be doing good when in fact they were creating misery. The need for labour in those huge establishments concentrated population in the areas around them. Cities became overcrowded and disease-ridden, water supplies

Children at work under oppressive overseers.

even more inadequate, builders unable to keep up with the demand for cheap housing.

'Some of the new works' it was said 'are most imposing structures, palatial in appearance, colossal in extent and durability, magnificence and comfort unsurpassed.' The houses the workers inhabited were filthy and overcrowded to the extent that even normally credulous commentators were driven to suggest that:

> employers of large numbers of work people who have brought to the town thousands of Irish and those of a lower class than have ever lived in it before owe it to the whole community to provide proper shelter for those they employ. The owners of large mills who have amassed fortunes and live in princely houses cannot permit their work people to live in wretchedness without having a heavy reckoning to pay in the future.

The wages in those palatial establishments remained very low and inevitably it was women and children who worked there.

> *Oh dear me, the mill runs fest,*
> *Puir wee shifters canna get a rest,*
> *Shiftin bobbins, coorse and fine,*
> *They fairly mak ye work for yer twa and nine.*

There are terribly affecting tales of men who could not themselves find work carrying their tiny children on their backs to the mills, the children dragged sleeping from their bed to work twelve and fourteen hours on end.

Because of the low wages spinners were looked down upon and even in a town like Dundee where both weaving and spinning were women's work, the spinners were thought socially inferior to the weavers. Weavers wore hats to their work and held their heads high. Spinners drew shawls over hair matted with mill dust and were thought low and coarse.

Spinning as a craft had sunk a long way since the days when the polished hand spinning wheel stood in every lady's room and its gentle whirring provided a background to family conversation.

7 Working conditions

It is hard, now, to understand how employers could have expected small children to labour for such long hours in their mills and factories. It is perhaps even harder to understand why their parents allowed it. The alternatives, however, were not attractive. For most girls there was only field work or domestic service. Work in the fields was just as hard as in the mills, the overseers just as brutal, the days just as long and cold wet weather made it worse. Girls as young as eight or nine were often sent into service away from home to work as skivvies, tweenies and scullery maids for mistresses who might or might not be good to them. For boys things might be even worse as orra loons, messenger boys, boot boys or chimney sweeps.

Eighteenth-century cottage interior - 'earth-floored, overcrowded, under-furnished dens'.

Even for those who stayed at home to work for their fathers as winders and reelers, life was not easy and there was little time for play. Their homes were not the rose-clad cottages of romantic stories but were more likely to be damp, earth-floored, overcrowded, under-furnished dens with leaking roofs and middens at their doors. When the mills were new they must indeed have seemed palaces of light compared to dark hovels and the idea of passing the time there in the company of other children, warm and dry, seemed, perhaps, at least to begin with, quite pleasant.

As competition caused owners to cut prices to the bone and the need to make a profit grew more and more pressing, the children were pushed harder and harder and conditions in the mills grew worse and worse. The giddy rate of industrialization and the overwhelming pace of town growth meant that the resultant disease and poverty became harder to ignore. People who were themselves comfortably placed began to be alarmed by the obvious degradation of the mill people, by their sickly appearance,

One of the Palaces of Light, *J P Coates's cotton-spinning mill, Paisley.*

the filth of their overcrowded streets and especially by the rate at which they died. Fear of disease combined with troubled humanitarianism to provoke consciences. Robert Owen was one of the first to cry halt and to press Parliament to legislate to protect child workers. His father-in-law David Dale, founder of New Lanark, had been praised for his benevolent treatment of his worker children who lived in 'well aired rooms, three in a bed' and while well fed and clothed, worked thirteen hours a day and then attended for another two hours at a factory school. Owen reduced the working hours to twelve and did not employ children younger than ten.

Unfortunately he was not successful in persuading his fellow mill owners to improve conditions in their mills. The mere suggestion that Parliament might make laws to restrict working hours brought the manufacturers together in protest. It would, they said, 'not only be unjust to those who had erected the mills on the faith of being unrestricted as to the hours of working, but bring distress upon the weavers by reducing the quantity of yarn that could be supplied to them'. But public opinion was beginning to fight against such greed.

Robert Owen's utopian views were well known and much distrusted by most manufacturers. He was thought of as a dangerous radical, likely to lead the country into ruin with his wild ideas. His proposed parliamentary bill was watered down until it was almost meaningless and could not be enforced. The Factory Act of 1819, which resulted from it, applied only to cotton mills, where the evil was seen to be greatest, prohibited the work of children under nine years of age and restricted the hours of those under sixteen to twelve hours a day. This still meant that a ten-year-old would work from six am to six pm and that he would go home to an empty house because the hours for adults were not restricted and his parents might be working for another three or four hours.

It took the power and authority and irreproachable respectability of Lord Shaftesbury, 'the Children's Friend', to push through the House of Commons the first and the most important

The Earl of Shaftesbury, known as 'the Children's Friend' because of his radical factory reforms.

of the acts to protect working children. The Factory Act of 1833 applied to all mills, not only the cotton mills. It forbade the employment of children under eighteen during the night. It forbade their employment for more than twelve hours in one day or 69 in one week. This was quickly reduced to nine hours for children under thirteen. It insisted on the attendance of mill children at school for at least two hours a day. It forbade the employment of children under nine. Children were not to be employed without a medical certificate which stated their age and passed them as fit for work. Unfortunately the mill owners were so eager for cheap labour, the parents so dependent on children's wages and, sad to say, medical practitioners so ready to turn a blind eye, that many under-age children still continued to work long hours in the mills. It was well known in Dundee, for instance, that certificates could be bought for sixpence.

But the most important thing about Shaftesbury's Act may have been the setting up of a system of factory inspection to see that the rules were properly carried out. The Factory Inspectors, with some blameworthy exceptions, did a good job. Their reports, detailing the conditions under which they found children working even after the Act, shocked Parliament and the voters and played an influential part in bringing in the series of acts which improved safety in the mills. They told of the unfenced machinery, the boiling water, the dangerous chemicals to which children were exposed. They reported the terrible accidents which scarred and crippled factory children and they eventually produced safer conditions for them.

There was still no thought of limiting the hours worked by adults. They were entirely dependent upon the demands of their master and could be made to work round the clock if he chose or if there were what employers called a 'press of orders'. Cases were reported of young women who worked 36 hours on end while tending one machine. They were required to watch it and to turn over the hanks of yarn at regular intervals. They would lie down on benches beside the machine and sleep for half an hour, then rise

Nelson's flagship HMS Victory.

The flax weavers of Dundee and Angus supplied Nelson's fleet with sail canvas.

Robert Owen, owner and manager of the New Lanark Mills, 1800-1825.

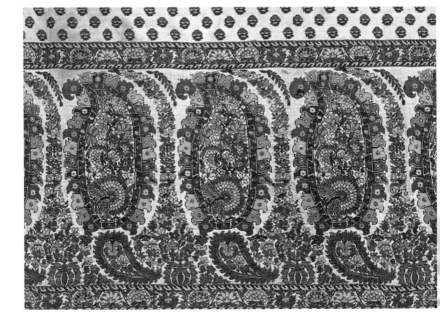

Left:

*Weaving an intricate pattern of
bird and leaves on a Jacquard
loom.*

Detail of a Paisley shawl pattern.

Above:

*Charles Campbell of Lochlane,
1760. The very model of a
Highland gentleman, though of
the Hanoverian persuasion.*

*Sample of naturally dyed wool on
a stick. The first eight colours are
derived from locally available
plants and lichens, the bottom two
from imported and greatly valued
indigo.*

to turn the yarn again for another spell. After 36 hours they would go home for a whole night's sleep, returning in the morning for another 36 hours. To make this particular case worse, the employers could see only one thing wrong with the arrangement and that was that young men and women were sometimes employed together at night and 'immorality might occur'.

Alarm about the health of young women in the mills and the effect on family life of their long hours at work eventually, in 1844, produced legislation to control the hours worked by females, which were reduced to twelve. In 1847 a ten-hour day for women and children was achieved. This still did not apply to men but, in practice, because in many mills men's and women's work was interdependent, men often had to stop when the women did. Men had to wait for protective legislation until they learned to organize in trade unions and build their strength by cooperation. The Act of 1874 eventually limited the work of adult male textile workers to ten hours per day.

The nineteenth-century Factory Acts are one of the great landmarks in human history. They were imperfect. There were many loop-holes. But they established the principle that profit should not be more important than human health and human safety. And they were first applied in Great Britain and in the spinning and weaving trades.

8 Primitive weaving

Prehistoric peoples are thought to have discovered the art of weaving when they first interleaved rushes, wattles or big-leaved plants like palm or banana to make shelters for themselves and their possessions. They also wove quite intricately contrived baskets out of reeds and twigs in which to gather berries and roots

Anne Campbell, Isle of Harris, one of the last users of the old handloom for tweed.

or carry flints and tools. The Ice Man, whose body was discovered high in the Alps and is thought to date from a period 5000 years ago, carried just such a basket and wore a cape woven from grasses.

Weaving simply means interlacing two threads, the warp and the weft. The weft is passed alternately between the threads of the warp. The first looms were very simple affairs, consisting only of a rough upright frame from which the warp threads hung on weights. The weavers passed a stick, around which the weft yarn was wound, under and over the warp. This stick was the very first *shuttle*.

Some North American Indians wove intricate hangings and blankets, and South American Indians still practise the weaving of very beautiful patterned cloth on simple narrow looms. The ancient Egyptians were capable of weaving very fine linens. Samples of textiles have been found in the tombs of the Pharaohs,

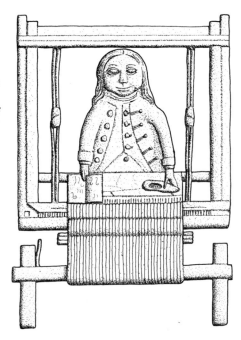

Weaver carved on eighteenth-century tombstone, Kirkton of Monikie, near Dundee. This stone was erected in 1765 by Edward Gibson. The good book at his side reads 'Faith and charity is good talents'.
Colin Hendry

and their wall paintings show the tall narrow looms on which they were made. Delicate muslins and silks were imported into Britain from the subcontinent of India long before weavers here could attempt work of such quality.

These ancient civilizations did not own looms of any greater complexity than those available in Scotland. In fact combs used for preparing fibre for the loom have been found at prehistoric sites in Scotland which are more or less the same as those used in ancient Egypt. The difference between the finest and the coarsest of textiles lay less in the machines on which they were woven than in the yarns of which they were made and the skill and experience of their users. Those skills were developed in answer to a demand for fine textiles which hardly existed in Scotland at the same period. The simple loom made of rough timber could provide what was required of it whether that was filmy silk or loose canvas. It remained largely unchanged throughout the world until the Middle Ages.

The origin of the horizontal loom is uncertain. A box-like framework with the warp, instead of being hung from weights as in the vertical loom, stretched between a *warp beam* at the back and a *breast beam* at the front. This allowed the first real improvement in weaving to be made. It became possible to divide the warp into odd and even threads and to fix these to horizontal sticks or bars called healds or, in Scotland, *heddles*. The weaver, by drawing the first heddle towards him, created a space between odd and even threads. So, instead of having to pass the stick or shuttle of yarn in and out slowly between the warp threads, he could now pass it straight through in one quick movement. Then by moving the second heddle the shuttle could just as easily be passed back between the alternating threads. By adding extra heddles to the loom different colours could be introduced and patterns invented. If the loom were narrow, as the early looms were, it was possible for the weaver to reach right across it to catch the shuttle at the other side. With broader looms it was necessary to have an assistant standing ready to catch the shuttle and pass it back.

This basic loom is still used in different parts of the world today. Until the Middle Ages it remained unchanged in principle and yet on that very simple, hand-operated mechanical device the most intricate skills were developed and the most beautiful cloth woven. The quality of a piece of cloth depends always on two things, the fineness of the yarn used and the imaginative skill of the weaver.

Of course minor adaptations and improvements took place all the time as weavers saw ways of making their own task easier and their product more interesting. By the time of the eighteenth century when both population and production were beginning to grow, the framework of the loom commonly in use in cottages and weaving sheds throughout Scotland looked something like a four-poster bed. It was known to some as 'the four stoops o misery' and the clanking of the loom and the flight of the shuttle were sounds which accompanied everyday life all over Scotland.

This loom was fitted with a roller at each end, between which were stretched the threads of the warp. Suspended from the top of the loom were two wooden heddle beams which could be raised by the weaver's foot on a treadle. Half of the warp threads, that is every alternate warp thread, were passed through eyes in the strings hanging from one heddle, the other half through the loops of the other. The weaver sat in front of the loom and by pushing down the right treadle with his right foot lowered one of the heddles, thus making a space, or *shed*, between the two sets of warp threads. The shuttle, wound with the weft thread, was thrown through this space, thus weaving the weft thread under and over the warp. This is the basic process of weaving and a loom of this sort can be seen in the Angus Folk Museum at Glamis.

The warp thread was commonly of a stronger yarn than the weft but all sorts of combinations could be introduced. The warp and weft could be of different colours, so producing what are called *shot* fabrics, the warp glinting through the weft in the finished cloth. Instead of fitting every alternate thread to the heddle it could be set to lift every two or three threads, or first two and then four, up to about a dozen, so creating striped and checked effects either in different colours or in self-coloured fabrics. From this practice all sorts of wonderful and complicated patterns were made possible even before the introduction of mechanical weaving. It was laborious and required great skill to weave the patterns seen in fine damasks, but it was possible.

The next important invention in weaving came in 1733 when John Kay invented the *flying shuttle*. Instead of the weaver passing

Drawing of handloom, showing the uses of the various parts. Note how the shed is raised by the heddle to allow the shuttle to be passed between warp and weft.

Damask, showing the intricate patterns which could be woven into the cloth by linen damask weavers even before the introduction of the Jacquard loom.

his shuttle by hand between warp and weft with an assistant to pass it back again, the shuttle could now shoot through on its own when he pulled a cord and shoot back again when he pulled another. This speeded production so much, some say it doubled the quantity of cloth produced, that weavers were afraid of being thrown out of work. In some places there were rioting and machine-breaking. But with an increase in trade demand increased and soon Kay's invention was very generally adopted.

In Scotland, just before Kay's patent, it was reported that 'the shuttles used by the weavers in this country were of a very coarse make and others of a better sort were ordered to be made after patterns brought from abroad'. And later, in 1745, notice was taken of 'an ingenious method of throwing the shuttle, invented by a native of this country'. In 1747 this inventor was awarded a prize of £70 for 'his new machine for throwing the shuttle'. Whether this was in fact an independent invention or a surreptitious copy of Kay's shuttle is difficult now to tell.

Most weavers until late in the eighteenth century, if they were not weaving for their own families, were occupied with *customer work*. They were using yarn spun by their customers or their customers' servants to weave goods on commission. The big houses of the neighbourhood would order tablecloths and napkins, striped stuff for petticoats, ticking for mattress covers. Mrs Mylne of Mylnfield, for instance, while holidaying in Bath, sent home to Perthshire samples of materials for bed hangings and dresses which were the latest fashion in the south and asked her household servants to have them made up by the local weavers. At the same time she reminded them to be sure to keep the maidservants spinning so that there would be enough yarn for new napkins and sheets when she came home.

9 The Board of Trustees and the weavers

The Board of Trustees was as anxious to nurture weaving as it was to improve spinning. The gentlemen who composed the Board were of a kind to delight in ingenious mechanical inventions and were always pleased to hear about new contrivances and machines and to reward their inventors.

They had surveys made of the state of the textile industry in Scotland and quite rightly judged that it was a poor, coarse trade. The mistake they made was in imagining that anything could change that until the Scottish economy improved. It is doubtful whether all the encouragement they gave and the money they spent made very much difference to the kind of cloth produced for sale during the century in which their influence was felt.

What they did was none the less interesting and the enthusiasm with which they approached their task can only be admired. They decided that it was necessary to introduce the Scots to a finer kind of cloth than they were used to making. They brought French Protestant weavers to settle in Edinburgh and to teach their skills. The part of France from which they came is remembered today in the street name Picardy Place.

They had a hard and miserable time here. The venture was unsuccessful for two main reasons, both of which might have been foreseen. The first was that the yarn with which they were supplied was wholly different from that which they had been used to at home. It was coarser than they were accustomed to, ill prepared and of a different twist, quite unsuited to the cambric weaving they had been brought here to demonstrate. The second reason for failure was that the end product, fine *cambric*, was too expensive for the Scottish market and perhaps unsuited to the

State of the Linen Manufacture from Christmas 1728 to Xmas 1729 continued

Of the value of 6ᵈ ℗ yard and under —————— 1.325.398 ⁴⁄₈

    above 6ˢ and not exceeding 1ˢ ℗ yard — 1.530.146 ⁵⁄₈

    above 1ˢ & not exceeding 1ˢ 6ˢ ℗ yard — 323.943

    above 1ˢ 6ˢ & not exceeding 2ˢ 6ˢ ℗ yard — 38.094 ²⁄₈

    above 2ˢ 6ˢ ℗ yard —————— 7.573 ⁴⁄₈

    Total as before ———— 3.225.155 ⁶⁄₈

The value whereof as per abstract Book is — £114383 „ 19 „ 8 ⁶⁄₁₂

And the medium price per yard is 8 ⁶⁄₁₂ ᵈ

The sum expended by the Trustees for promoting the Linen Manufacture this year was as under vizt.

For premiums for sowing Lintseed ———— £121. 10. —

For rewarding the Inventor of the machine for breaking and dressing of Flax and for defraying his expences ———————— 230. —. —

For models of Instruments for improving the Manufacture ———————— 39 „ „ —

For setting up seven Spinning schools ——— 150. —. —

For prizes on the best Webs of Linen Cloth — 228. —. —

For salaries to fifty Stampmasters ——— 500. —. —

For salaries to two General Riding Officers — 250. —. —

    In whole ——— £1619. 3. 9

Scottish climate, as indeed were the poor Picardy weavers who complained bitterly of the cold and damp and darkness.

Holland had, at the beginning of the eighteenth century, the best reputation for linen production in Europe. The Trustees encouraged travellers in Holland to study their methods secretly (they were certain that there was a secret) and to report back what they found. This was in fact a case of industrial spying. They also brought Dutch weavers and Dutch looms over to Scotland so that the Scots weavers might learn from them. They paid the salaries of a number of Dutch master weavers for some fifteen years and these men travelled throughout the Lowlands of Scotland, exhorting and instructing those Scots who were willing to be taught. They did have some success in spreading knowledge of the best methods across the country, and both the quality and the quantity of fine linens did increase over the period in which these Dutchmen were riding the Scottish roads. By 1744 the Trustees could claim in their yearly report that:

> The looms and takle for weaving of table linen were now so much improved in this country both by means of the foreigner some years brought over here, and likewise from the ingenuity of our own workmen who did even very much improve on his discoveries. Some apprentices had been bred under his direction at the publick expense and being now fully masters of the business the Trustees proposed to give them some of the best kinds of looms and other takle for carrying on their business both for enabling them to carry it on even better, and to serve as patterns to be imitated by other ingenious tradesmen. And for saving expense to the publick the foreigner was permitted to return to his own country.

We cannot tell whether the increase and improvement of fine linens would have taken place without the Trustees' help or not. Certainly a few individuals benefited from their willingness to

Minutes of the Board of Trustees, 1728/9.

help with setting up small weaving businesses. What is certain, however, is that, while the quantity of fine linens did increase, it did not increase at anything like the pace of the coarser products, the soldiers' sarking, the sail canvas and sackings and the rough clothing for the slaves in the American plantations.

The heavy-linen manufacturers, although their skill was less spectacular than that of the damask weavers, were depended upon for everything requiring strength and durability. The sail canvas weavers of the east coast provided for the Royal Navy and supplied the sails for the Fleet at Trafalgar. Baxter Brothers of Dundee, the largest flax-weaving firm in the world, produced tarpaulins for the covered wagons which pushed west to settle the American Plains. Later on, Macgregor the hose-pipe manufacturer of Dundee was one of the last firms to employ handloom weavers. They were kept busy supplying a huge demand for linen canvas hose pipes created in New York by the building of skyscrapers. The coarse-linen trade proved more enduring in the end than might have been expected.

Tarpaulin for the covered wagons which carried emigrants across the American continent was supplied by the canvas weavers of Dundee.

The most important achievement of the Board of Trustees was the setting up and administering of a system of inspection and control. They placed salaried stamp-masters in every population centre in Scotland and gave them the power to reject goods which did not reach a set standard of respectability. They judged colour, strength, closeness of weave and general appearance. They insisted upon honest measurement and would unfold a web to make sure that it was of a proper length and did not contain faults folded inside it. They employed assistants known as *lappers* to help them unfold and fold, or *lap*, the cloth. When they passed it as satisfactory they stamped each length with the Trustees' stamp and the stamp-master's own name, the name of the county in which he operated and the length and breadth of the cloth upon both ends of the piece. 'And such cloth as they found not sufficient they were directed to cut into unmarketable lengths and to seize all linen cloth that should be offered to be sold or packed up for sale without being stamped.' One crafty weaver is said to have got over this by cutting all the rejected pieces into the shape of soles and selling them to a shoe manufacturer.

Without the stamp the goods could hardly be sold at market in Scotland and, more importantly, were not eligible for the government bounty on exported cloth. By this means the Trustees raised the reputation of Scottish goods both at home and in overseas markets and established a standard of good practice very much in advance of its time.

A few of these stamp-masters turned out to be rogues. They were, after all, in a position which left them very open to the temptation of bribery. A few of those in the areas nearest to the Highlands were suspected of being 'disaffected from His Majesty's person', that is they were supporters of Prince Charlie. But most of them seem to have been honest and hard-working officials and some of them earned a position of reputation and respectability within the neighbourhood of their stamp offices.

The Board of Trustees, of course, supposed that the spinners and weavers of Scotland would be occupied, as traditionally they always had been, in using wool and flax, fibres produced in their own country. Towards the end of the Trustees' period of influence, however, there were changes.

Cotton, tended and harvested by black slaves on the plantations of the southern states of America, began to be imported into Glasgow where trade across the Atlantic was already well established. It was cheaper and easier to work and soon supplanted flax in the western half of Scotland.

In the eastern counties, where the coarse linen trade continued to grow in bursts of prosperity interrupted by periods of depression, experiments in keeping down costs led to the use of foreign fibres. First hemp and then jute were imported into east-coast ports from India where the East India Company was keen

Jute warehouse in Dundee, with horse and cart waiting to carry bales of raw jute. The cobbled streets of Dundee were set with smooth granite tracks to assist the passage of laden carts from dock to warehouse to mill.

to use contacts in Scotland to capture new markets for Indian products.

Hemp is a plant of the nettle family. The English word *canvas* comes from *cannabis*, the Latin name of the plant, whose leaves are used in the manufacture of the drug cannabis. Hemp was widely used in rope and sail-making although the Board of Trustees frowned upon its use for sail canvas for the Navy because it could not produce a material as durable as flax canvas.

Jute is made from the bark of a tree-like plant and, because it was cultivated with low-cost Indian labour, it could be imported into Scotland very cheaply indeed. Its use gave the textile manufacturers some problems but, as well as its cheapness, it had the advantage of taking dye colours easily and before too long it

Camperdown Works, the creation of the Cox brothers. The biggest jute works in the world, now mostly demolished.

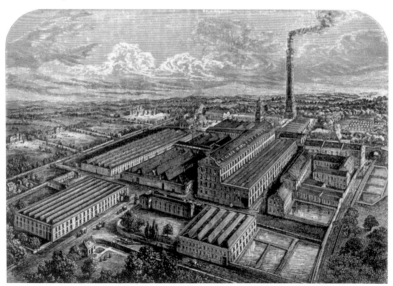

replaced flax at the cheap end of the market, especially for sackings and floor coverings.

Those who first attempted to use jute as a substitute for flax found it difficult. It was a very dry, brittle fibre, liable to snap under pressure and therefore difficult to adapt to machine use. In jute factories the preparatory processes took up acres of floor space with heavy machinery. Thus where the cotton mills, with their preponderance of light spinning frames, could be built high, the jute mills spread across a greater area of ground in lower buildings. Their builders' wish for height and grandeur was satisfied by the erection of enormously high chimney stacks and elaborate bell towers.

Mill chimneys, Lochee Road, Dundee. The air
over the town used to hang black with smoke. SEA

The machines broke and rolled and combed the raw material until it changed from a dark, rough, woody bundle into soft golden fibre of the texture of fine human hair. To make all that machinery work without constant stoppages it was found necessary to soften the fibre by soaking and spraying it with oil. As the most suitable oil available for the purpose proved to be whale oil, whose chief use before that had been in lamp lighting, this brought a boost to another Dundee industry, the hunting of whales. Almost every linen manufacturer in Dundee attempted the weaving of jute, at first in conjunction with a linen warp thread. Even the important Baxter Brothers tried it, although in the end they abandoned the experiment and decided to stand almost alone as a flax firm.

Although jute was introduced into Dundee in the 1820s and, after many experiments, spun with uncertain success from about 1832, power-weaving of jute was not attempted on any scale until the middle of the century. Hand-woven jute competed with power-loom linen until the establishment of Cox's factory at Camperdown in 1849 and Gilroy's in 1851.

Because of the many uses to which their products were put by the armed forces the jute and coarse-linen trades experienced their best years during wars or periods of threatened war. The tendency to take advantage of seasons of quick, easy profits for the over-extension of the industry while demand lasted brought severe recessions between wars.

The problem with using a fibre imported from India, however cheaply it could be got, was that the products made from it would always have to compete with articles of the same kind produced in its country of origin where wages were very low and workers used to poor conditions. To beat Indian products in world markets Dundee manufacturers had to keep their own costs very low while at the same time aiming for superior quality and durability. This could only be done by manufacturing on a very large scale so that tiny profits were multiplied by enormous turnover into worthwhile takings. It meant employing thousands of workers at very

'The Pletties', overcrowded tenements in Dundee. Access to the upper flats was by way of outside stairs and stone platforms (pletties). The later brick addition housed communal toilets, an improvement on the earlier midden on the back green. SEA

low wages and when the industry expanded to the extent that there were not enough workers available in the textile towns of Angus, it meant attracting thousands of workers from Ireland to come and live and work in Dundee. House building could not keep pace and the town became seriously overcrowded with an underpaid, ill-nourished, disease-prone population. Jute brought fortunes and ostentatious mansions to a few families, poverty and slums to the thousands who came to depend on them.

10 Good days for the handloom weavers

Good days for the handloom weaver came in the period between the establishment of the first spinning mills and before the introduction of powered weaving. These were what one old manufacturer called 'the daisy days of weaving'. Traditionally his problem had been to keep a steady supply of yarn. But improved farming, better heckling and bleaching and the vastly increased rate of production in the new mills began to result in a dependable supply. With plentiful flax yarn at reasonable prices, with looms improved by the flying shuttle so that they were much less laborious to use, and with an improving economy which kept orders and commissions flowing in, the weaver was, for a time at least, in a strong position.

He had a money income, enough to feed and clothe his family and a little over. Observers began to notice the appearance of luxuries like clocks and tea kettles on cottage mantelpieces. He had steady employment, a position in society and a skill which he could practise with some pleasure if he had an interest in improving design and introducing different types of weave.

Skill in design was encouraged by the invention of the wonderful Jacquard loom. This machine, a forerunner of the computer, used punched cards to control the weft on the loom and made possible the weaving of very beautiful and intricate patterns. Dunfermline was already the centre of the fine table-linen industry and its damask weavers were able, even before Jacquard, to weave complex patterns bearing coats of arms, heraldic devices and family mottoes into their tablecloths. With the new machine no pattern became too complicated or intricate for the Dunfermline weavers and their work, particularly their beautiful flower patterns, became widely known and desirable all over the world.

A handloom of this sort can be seen at work in the courtyard of the National Trust of Scotland property, the House of Dun, near Montrose, where the last handloom linen weaver in Scotland carries on his business.

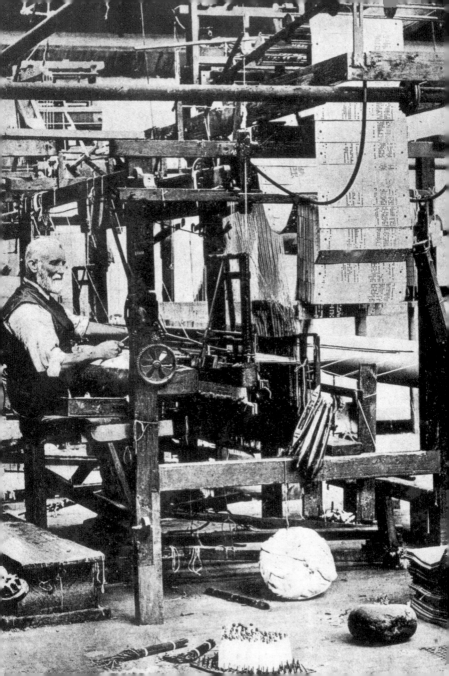

Some weavers, enjoying their prosperity, began to take on apprentices, purchase extra looms, employ other men to work for them, perhaps half a dozen of them in a weaving shed. They began to call themselves *manufacturers* rather than working weavers. It was from this class that, eventually, many of the large-scale manufacturers of the nineteenth century arose. Having in the days of good trade put by a little money to give them working capital, they were in a position to invest in the new machinery as it was invented, and to buy and sell in the markets with advantage.

The handloom had by now reached a kind of smoothness and effectiveness which could not easily be improved upon by mechanical operation. For this reason and because the application of power proved very difficult and complicated, weaving for the most part moved straight from hand to steam without the intermediary step of water power. It was not so much any wish to improve the product or the working of the machinery that was behind the mechanization of weaving as the need of capitalists to organize on a large scale.

11 Power weaving

Power weaving, when it came, ended the good days for the individual handloom weavers, but it came at different times for the different branches of the textile trade and spread unevenly over the country. So, while cotton, as with spinning, was quickly and easily adapted to machine weaving, linen-weaving machinery was at first unsatisfactory. As late as 1838 there were still 51,060 handlooms working in the Lowlands of Scotland of which 2,400 were linen and woollen looms, the rest occupied with cotton.

Edmund Cartwright patented the first power loom in 1787. He was a clergyman, with no mechanical knowledge whatever, but

Weaver using a Jacquard handloom. The punched cards which controlled the complicated pattern-making hang to the right of the picture. SEA

59

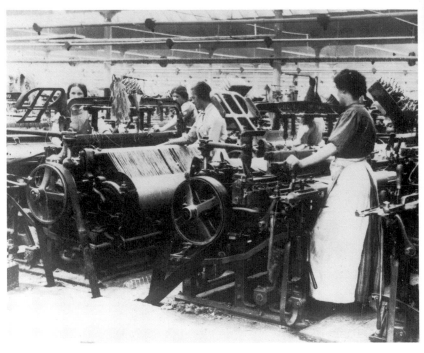

Weaving shed, Hillside jute works,
Dundee, 1911.

he was struck by the notion that power might be applied to the process of weaving and he succeeded, with the help of a village carpenter, in building a powered loom which was quickly taken up in Manchester. Attempts were made to adapt the machine for linen but the first firm in Britain to succeed was the London firm of Charles Turner and Company in 1813. Power-loom linen weaving was attempted in Dundee in 1828 but it was not found possible to do it profitably because of the endless breakages.

In 1824 the great financier John Maberley set up the first successful powered linen weaving factory in Scotland at Aberdeen with two hundred looms. When asked what was the motive for his

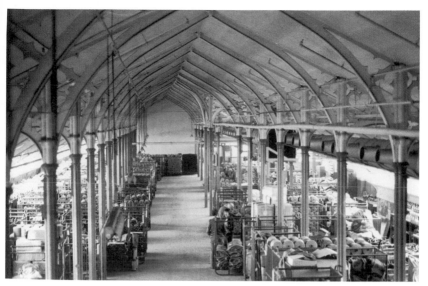

*Camperdown Works, Lochee, showing the light
and airy cast-iron structure.* Bruce Walker, SEA

experiment he answered in the House of Commons 'Profit, of
course!' And when he ceased to find the venture profitable he ret-
urned to London leaving his Aberdeen co-partners in difficulties.

In 1836 Baxter Brothers of Dundee set up their first power-
loom factory. It looked very fine. The weaving room was 150 feet
long and 75 feet wide, the roof was supported by cast-iron pillars
and lit by roof lights. It contained 216 looms and employed 300 or
400 people. But it disappointed its owners.

> Each loom had an individuality of its own and the idiosyncrasies of
> each used to be freely pointed out and commented. Of one it was
> said that she would only do a certain kind of work, of another that
> she had taken a stubborn fit and had to be coaxed for half a day
> before she would throw the shuttle or take up. Being so fitful and
> apt to get out of order, the looms, as might be expected, did not
> produce much work.

It was not until after 1825 that mechanical cotton weaving was successfully carried out and into the 1840s before the machines introduced for linen were found to be profitable. But then the handloom weavers were doomed. They were forced to leave their cottages, seek employment in the factories, work for a master and lose their independence. Most of them found it very hard. Where they had been accustomed to take a day off now and again, now they must attend long regular hours or risk losing their jobs. Where they had bowed to no one, now they must accept discipline and kowtow to employers. Worst of all, they were to see the job which had been a source of male pride taken from them by girls, the handloom which had required the whole of their attention now replaced by a row of looms which could be supervised by one young woman.

Some branches of the trade survived. Those who were specialists, who made very fine high quality goods individually designed for an individual customer, those who could do something still too difficult to be achieved on a machine or which had a limited but prosperous clientele, these found a niche in the industrializing society and survived within it. And those who lived at a very great distance from the cities and could work for a restricted community found a way to make a living.

12 Paisley shawls

The weavers of Paisley were said to be the most intelligent and the most widely read of all Scottish workers. In 1766 there were 1,767 handloom weavers in the town. After the introduction of shawl manufacture in 1803 the number grew to 7,000.

The shawl was a fashion accessory, essential to cover the thin, clinging muslin dresses which became fashionable at the end of the eighteenth century. At first shawls, made of the extremely fine and durable wool of the Kashmiri goat, were imported into this country from India. They were very expensive but proved to be so popular that home weavers quickly

These mid-nineteenth-century travellers show the usefulness of shawls when tailored coats for women were not commonly available. The older woman wears a large Paisley-patterned shawl folded about her shoulders. The young man wears a plaid of shepherd check.

attempted imitations. Although the first shawl manufacturers began in Edinburgh and in Norwich, Paisley soon became the centre for a very important and prosperous trade. From 1805 until about 1870, when the fashion passed and ladies' tailored coats and jackets replaced shawl wearing, Paisley weavers were occupied with the making of shawls in silk gauze, muslin and fine wool of wonderful colour and pattern. Paisley had already, in the eighteenth century, established a reputation for silk weaving. In 1781, out of 6,800 handlooms in the whole district, 2,000 were weaving linen and 4,800 silk. The Paisley weavers had the skill to take advantage of the very demanding new

*Handloom weaver at work with his draw boy. The
boy (wearing a glengarry) pulled the cords at the
master weaver's command, and in this way produced
intricate patterning even before the introduction of
Jacquard looms - though much more laboriously.*

fashion when it came and they were already equipped with
looms suited to the needs of the industry.

The draw loom was a heavier and more complicated piece of
machinery than the ordinary handloom. Its apparatus of lead
weights, cords and harness was difficult to set up and required
skill and strength to operate. Each weaver needed the assistance of
a draw boy, who had to pull the different bundles of cords which
controlled the pattern-making in answer to the weaver's call.

The ambition of the trade was to lower the cost of making
these intricate and beautiful patterns but the weavers took pride in
creating new designs and in satisfying the particular requirements

of their customers. Paisley shawls were often given as wedding presents and a specially treasured shawl was traditionally worn to the christening of a new baby and known as a *kirking shawl*. Of many patterns the *pine cone*, which was an attempt to imitate the original Kashmiri pattern, remained the most popular and is the

Kirking shawl, worn on a woman's first visit to church after a wedding or a christening.

one which is everywhere recognized as *Paisley pattern* today. Most of the patterns were based on stylized flower designs. One very interesting pattern often seen has a border of formally styled carnations or pinks because the Paisley weavers shared a particular hobby of growing competition pinks.

The trade paid well enough and was sufficiently highly regarded to give the handloom weavers of Paisley wages high enough to allow them some spare time for pastimes such as gardening. Although machine shawls were manufactured, and printed cotton Paisleys became popular at the bottom end of the market, they did not at first greatly endanger the market for high quality hand-made shawls.

After the introduction of the Jacquard loom the domestic weaver did find it hard to compete. The weaving process itself was very much speeded up. On the draw loom it took something like two weeks to weave a shawl. On the Jacquard it could be done in one day. The setting-up process on the handloom was very tedious and time-consuming. On the Jacquard, patterns could be changed in minutes. But the machines were expensive and too big to be installed in cottages. They made inevitable the move into factories and they made impossible the commissioning of individual designs for individual customers.

It was not only mechanization that finally impoverished the Paisley shawl weavers and drove them reluctantly into the factories, it was the fading of the fashion on which their prosperity and independence had been built.

13 Tartans

Handloom weaving of woollen cloth was another survivor. Although the origins of a checked weave are very ancient, the brightly coloured *tartan* cloth with which Scotland is today so widely identified does not, in fact, go so far back in history as is popularly supposed. The notion that particular families had a 'right' to certain patterns is almost certainly a nineteenth-century

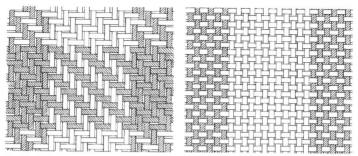

*Twill (left) and plain weaves show how differ-
ent weaves and colours make possible a wide
variation of patterning.*

invention. It suited the manufacturers to fall in with romantic
ideas about the habits of the Highlanders. These ideas were
highly exportable and good for trade even if founded in fiction.

Something resembling the kind of cloth we commonly know
as tartan dates from about the middle of the seventeenth century.
It was a hard cloth, with a twill weave. The pattern was the same
in warp and weft. It was made from a smooth, tight-spun woollen
yarn and was shrunk after weaving which is what gave it its hard,
dense quality. While *plain* weave involves taking the shuttle
between each alternate warp thread, for *twill* weave the shuttle
takes up two warp threads at a time instead of one. This gives the
cloth an extra firm surface.

Because Highlanders used chiefly dyestuffs such as berries
and lichens they could dye only small quantities at a time and it is
thought this may be the reason why, instead of weaving whole
webs of one colour, they chose to weave in checks and stripes of
different colours. A pattern which can be easily remembered
makes the weaver's task simpler and so small blocks of colour
repeated across the width of the cloth are to his advantage. The
Highland web was narrower than was customary in the Lowlands
and the flying shuttle, which was of such benefit to the weaver of
broad linen webs, was of less use to him: it is likely that Kay's

invention was not used by the handloom tartan weavers. The extra speed given by Kay's shuttle could not in any case greatly change the course of tartan weaving which was of necessity slow and laborious.

The idea that particular patterns were restricted to particular clans may have derived from a lack of ambition on the part of the local weaver rather than any clan loyalty or special wish to iden-

Designer's plan for the London Caledonian tartan, with sample. SEA

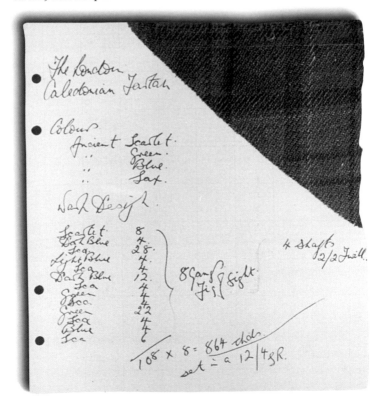

tify clan members. Once they had learnt a pattern they were inclined to stick to it rather than make their lives difficult by learning another one. Because they worked alone and at some distance from one another their patterns might indeed remain for a long time their own, but they were spread by other weavers passing through, admiring a particular check and perhaps adapting it by adding one stripe of another colour, or by visiting customers bringing a piece of cloth to be copied.

The defeat of the clans in 1745, and the laws which forbade the wearing of Highland dress, in fact had the effect of interesting a wider public in tartan cloth. When in 1782 the laws were repealed, those with romantic Jacobite leanings, quite safe now from accusations of treason, were inclined to adopt the wearing of it. Later, with the help of Sir Walter Scott and the Royal Family, tartan became seriously fashionable.

It was not until after mechanization that the idea of clan tartans with a right or privilege attached to them was introduced. It was a brilliant marketing ploy on the part of the nineteenth-century capitalist tartan manufacturers rather than a genuine tradition but it certainly worked. It sold, and continues to sell, tartan cloth in the commercial market in a way nothing else could have done.

14 Tweed

Tartan, however, was not the only Scottish contribution to the history of woollen textile design. One of the things which affected the kind of material woven in a particular area was the kind of yarn available. In Scotland the native sheep produced a mixture of dark and light wool. David Wedderburn, for instance, the six-teenth-century Dundee shipmaster whose records have been published by the Scottish History Society, shows purchases of wool from the Angus glens, some white wool and, at a lower price, some black. In the Borders, tradition has it that, rather than go to the expense of dyeing, the weavers sorted the blacks and whites

into two shades to weave the *shepherd checks*. This distinctive pattern of small black and white squares was worn at first by herds and shepherds and later became very popular with the gentry.

The word *tweed* to describe a special kind of Scottish woollen cloth is now so very generally known that it is quite difficult to accept that it was unknown until about 1830, was applied within the factory weaving industry and was not used in the traditional hand-weaving craft before that date. The story is that a clerk in a London firm misheard the word *twill*, locally pronounced *tweel*, perhaps confused it with the name of the river Tweed which runs through the Border country, and wrote down tweed in his order book. The name caught on, proved a good seller and has been applied to Scottish woollens ever since. It has come to be accepted all over the world as a description of hard-wearing, pleasantly coloured woollen cloth.

What has perhaps not often enough been commented upon is the remarkable contribution that good design has played in creating custom for Scottish manufactured tweeds. Design does not only mean applied pattern. In the case of tweed it applies to the very clever combinations of gentle, subtle colours which are woven into warp and weft of the varying makes of tweed. The Border mills, in particular, have acquired a worldwide reputation for the brilliance of their designs, their understanding that it was necessary not only to impress customers with the quality of their goods but also to capture the fancy of the market for novelty and beauty.

One present-day survivor of the handloom weaver's craft is *Harris tweed*, still woven in the traditional way, though with some adaptations, in the Western Isles. Harris tweed is notable not so much for its design, although its colours are distinctive, as for its

A Perthshire shepherd wearing a shepherd's plaid, a large shawl of black and white checked wool, useful for protecting both the shepherd and his newborn lambs. Sometimes the plaid has one corner turned back for use either as a hood for the shepherd or a pocket for the lamb. SEA

*Cheviot and blackface sheep, the main breeds
for wool production in Scotland.*

hard-wearing texture. Its success seems to be in part due to its
being granted its own special trademark so that it cannot be imi-
tated but is immediately recognized as a mark of quality worldwide.

The character of Harris tweed is given by the wool of the locally
raised blackface sheep just as the Border tweeds get their soft,
firm, thornproof quality from the Cheviot, a longer and bigger
sheep than the little blackface. Blackface wool is coarse, even
harsh, with some roughness or imperfection to give the charac-
teristic knopped effect, and it takes colour well to give the strong
gingery or bright blue hues of the Harris fabric. Nowadays, while
blackface is the typical Harris sheep, a small flock of Cheviots
grazes on Pabbay, an island in the Sound of Harris, and both
fleeces are used by the local weavers and spinners. While almost,
but not quite all, the yarn is now mill-spun, the hand weavers of
the Island of Harris have become one of the tourist attractions of
Scotland and their cloth an important item of export.

Most Harris tweed weavers now work on Hattersley treadle
looms in weaving sheds containing several looms, but there is at
least one working at home using the old *beart mhòr*, a handloom
with four foot pedals and a shuttle thrown by hand. The Gaelic
name for what we call Harris tweed is the *Clò Mòr*, the 'Big

Cloth', which was dyed with native plants, ragwort, heather, peat and *crotal*, the grey lichen which clings to rocks in the islands. It was waulked by hand, that is washed in hot soapy water then beaten to shrink and firm it into the tight-woven, weatherproof finish for which it is prized. Waulking used to be a social occasion, with all the women of the district gathering round when a web was finished and cut from the loom. They

Preparing a handloom for woollen cloth weaving in the Outer Hebrides, 1925. SEA

Loading wool, Skye, 1922. The dinghy is on the shore and the steamer lying at anchor. SEA

would sit around the table on which the wet web was laid, pounding and turning the cloth in time to the rhythm of traditional Gaelic waulking songs.

There are places, then, where wool manufacture has endured in something like its old, traditional form. But fine linen has almost disappeared and coarse linen and jute have been almost entirely superseded by polypropylene. Whole rooms full of machines can be tended by one person and activated by the throw of a switch. The use of man-made fibres has allowed the rescue of an industry and the endurance of some old textile towns like

> *The almost magical production of polypropelene.*
> *Polypropelene is extruded rather than spun. Its machines*
> *need very few attendants and so it represents the end of an*
> *old story for Scottish spinners, while making survival possible*
> *for one part of the Scottish textile industry.*

Forfar, the continued use of some old factories and factory sites and the survival of some old family names.

Scottish spinning and weaving began as low-skill crafts using poor-quality materials in a poor country, under-capitalized and overpowered by a stronger neighbour. After industrialization they competed in a fierce market by lowering prices and quality, only to succumb to foreign competitors.

But in modern times they compete on quality and skill and have managed to acquire and maintain a niche in world markets, their suitings and their tweeds in demand by the most prestigious tailoring establishments.

FURTHER READING

BAINES, Patricia *Spinning Wheels, Spinners and Spinning*, London 1977

BROWN, William *Early Days in a Dundee Mill*, ed John Hume, Abertay Historical Society 1980

BUTT, John and PONTING, Ken *Scottish Textile History*, Aberdeen 1987

CATLING, Harold *The Spinning Mule*, Newton Abbott 1970

CHEAPE, Hugh *Tartan: The Highland Habit*, Edinburgh 1992

DURIE, Alastair J *The Scottish Linen Industry in the Eighteenth Century*, Edinburgh 1979

GAULDIE, Enid *The Dundee Textile Industry, 1790-1885*, Scottish History Society 1969

HAMILTON, Henry *An Economic History of Scotland in the Eighteenth Century*, Oxford 1863

HORNER, John *The Linen Trade in Europe during the Spinning Wheel Period*, Belfast 1920

LEADBETTER, Eliza *Handspinning*, London 1976

PIKE, E Royston *Human Documents of the Industrial Revolution*, London 1966

REILLY, Valerie *The Paisley Shawl*, Glasgow 1992

SMOUT, T C *A History of the Scottish People, 1560-1830*, London 1969

WARDEN, A J *The Linen Trade, Ancient and Modern*, Dundee 1864

WATSON, Mark *Jute and Flax Mills in Dundee*, Tayport, Fife 1990

PLACES TO VISIT

Brodick, Arran: Isle of Arran Heritage Museum includes weaving material and occasional weaving demonstrations.

Crieff: the Highland Tryst Museum has a working handloom weaver.

Dalmellington, Ayr: Cathcartson Visitor Centre: a row of eighteenth-century weavers' cottages which feature a loom and other textile material.

Dundee: McManus Galleries feature material on the Dundee jute industry.

Dundee: Verdant Mill, a textile museum housing jute machinery and historical displays, opens in 1995.

Dunfermline District Museum features a collection of linen damask.

Edinburgh: the Museum of Scotland, due to open in 1998, will feature prehistoric spinning and weaving material and a major display on the Scottish textile industry.

Edinburgh: Scottish Agricultural Museum, Ingliston: includes material related to spinning and handloom weaving.

Galashiels Museum and Exhibition, includes artefacts and early photographs of the woollen trade, and a working loom run by a water turbine wheel.

Glamis, Angus: Angus Folk Museum

Harray, Orkney: Orkney Farm and Folk Museum has a weaving loom.

Hawick Museum has displays on knitwear and hosiery.

Spinning with distaff and spindle. An aquatint by D Allan, 1788. SEA

Inveraray, Argyll: Auchindrain Museum of Country Life.

Irvine: Glasgow Vennel Museum and Burns's Heckling Shop has a display related to Burns's time as a flax dresser.

Irvine: Irvine Burns Club Museum and Burgh of Irvine Museum has a room devoted to Burns's work as a flax dresser.

Kilbarchan, Renfrewshire: weaver's cottage (National Trust for Scotland) with eighteenth-century handlooms and weaving equipment.

Kingussie: the Highland Folk Museum includes spinning and weaving material.

Kirkintilloch: Barony Chambers Museum has textile photographs and equipment.

Montrose: Scotland's last handloom linen weaver can be found at the House of Dun (National Trust for Scotland) near Montrose.

New Abbey, near Dumfries: Shambellie House Museum of Costume features changing displays of European costume, including Scottish material.

New Lanark: restored eighteenth-century cotton-spinning village founded by David Dale and Robert Owen, with displays and working machinery.

Newton Stewart Museum includes costume and lace.

Paisley Museum and Art Gallery with strong displays on the local weaving community, including a history of the Paisley shawl.

Portree, Skye: the Museum of Island Life includes a weaver's house.

Stirling: the Scottish Tartans Museum features tartans, Highland dress and weaving material.

Strathaven: the John Hastie Museum includes artefacts on local handloom weaving.